It was a bright day in
Jellybean Town. Everyone was
going down High Street. The king
was there on High Street too.

Suddenly, there was a loud sound. "What could that be?" said the king.

A mighty critter was yelling in the sunlight. He stood on a hill near the town.

"We must do something fast," said the king to his knights.

Soon a brave knight made a very big mouse trap. He set the trap high on the hill.

Then the mighty critter looked at the trap. He said, "I'm a mighty critter, not a mouse!"

The mighty critter kept on yelling.

"What can we do to stop him?" said the king.

"You might have a word with him," one knight said.

The king met the mighty critter. He said, "Dwight, why do you yell?"

"A pebble is stuck in my foot," answered Dwight. So the king pulled it right out!

After thanking the king, Dwight went right home. That night he read by candlelight. And he did not yell again.

The End